We hope you
enjoy the book!

Water is precious. It is the source of life for everything on this planet.

2
Condensation
Vapor rises, cools, and turns into tiny water droplets to form clouds.
Clouds
1
Evaporation
Sun heats the ocean water and turns it into vapor.
Rain
3
Precipitation
When droplets get too heavy, they fall to Earth as rain.
Mountains
Lake
River
Ocean
4
Collection
Water collects in rivers, lakes, and the ocean. The cycle begins again!

Water never stays still. It travels all around the world in something called the water cycle, moving from the deepest oceans all the way up to the tallest mountains.

Water can change shape.
It can be a soft cloud,
a solid piece of ice
or a flowing river.
Our story follows
a little droplet
named Aqua.
Who is about to
discover the magic
about the water cycle.

One sunny morning,
Aqua woke up under
the glow of a bright and
beautiful rainbow.
He looked around
feeling confused.
Where was he?
How did he get here?
Everything sparkled and
shimmered around him.

WELCOME TO
RAINBOWVILLE

Aqua had landed in a wonderful place called Rainbowville. A magical town that only appears when sunlight shines through raindrops. It was colorful, cheerful and full of friendly faces. As Aqua walked through town a kind and friendly droplet came to greet him.

It was Mayor Dewdrop! "Good morning little one," said the mayor with a warm smile. "You look like you have lots of questions."

"I do!" said Aqua. "I woke up here and I don't know why... and why am I so clear and shiny?"

Mayor Dewdrop chuckled softly and said, “come with me, I think I can help.”

“This happens a lot here,” said Mayor Dewdrop kindly.

"New droplets arrive all the time without any memories, but don't worry everyone in Rainbowville looks after each other. You are safe here."

They walked to the towns center where a tall colorful clock tower stood.

“This clock is very important,” said Mayor Dewdrop. “At midnight Rainbowville goes to sleep until the next Rainbow appears. So, make the most of every moment.”

Aqua starred at the clock.

"Tick, tock... tick, tock."

Even though Rainbowville was so beautiful he could feel time moving.

He wanted to use every second wisely and find out where he truly belonged.

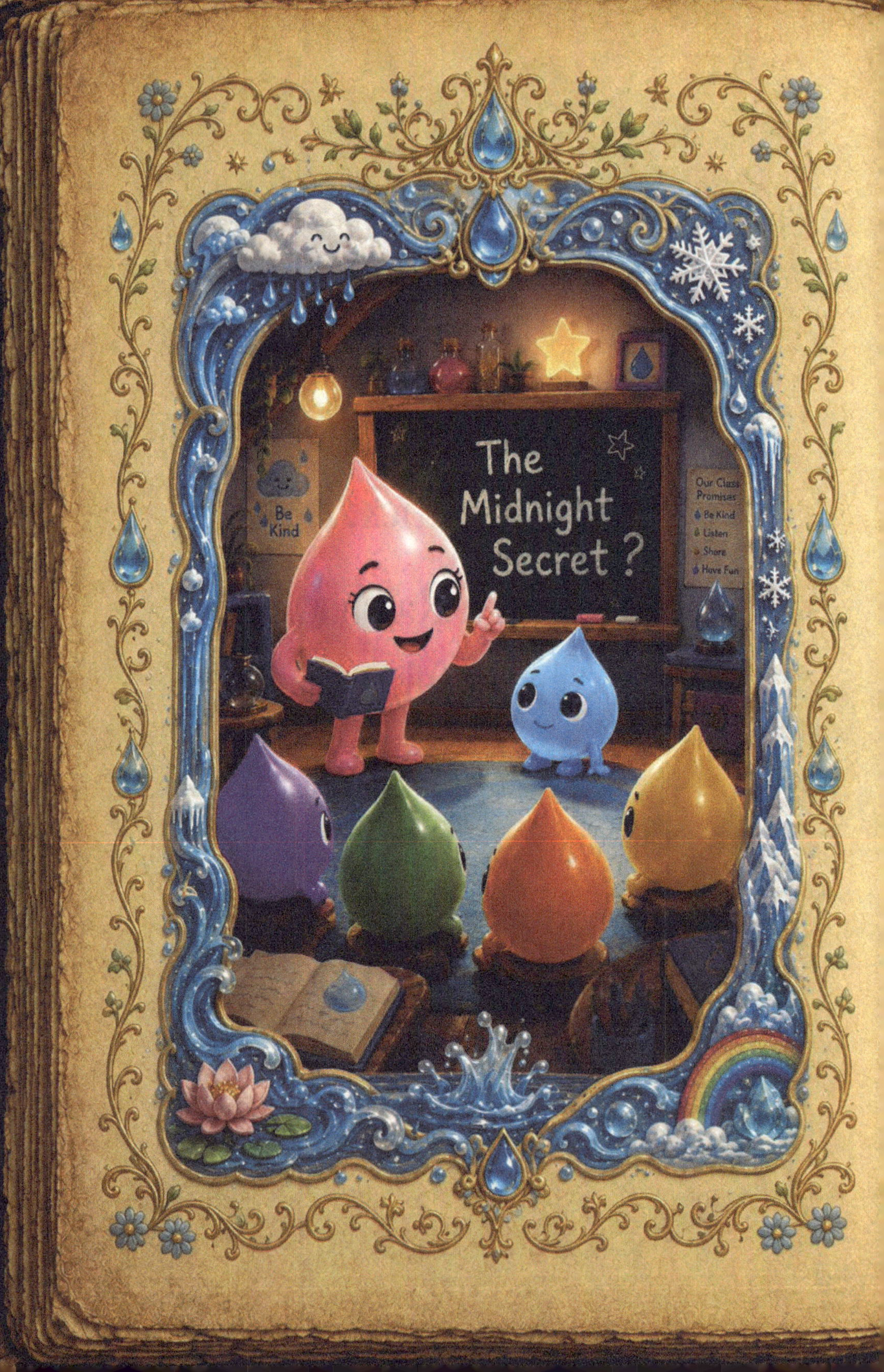
The Midnight Secret ?
Be Kind
Our Class Promises
Be Kind
Listen
Share
Have Fun

Aqua met a cheerful droplet named Misty who was a teacher in Rainbowville.

"What happens if someone is still awake at midnight," asked Aqua curiously.

Misty smiled mysteriously and said, "that is one of Rainbowville's greatest secrets."

"Those still awake at midnight," Misty whispered, "Become part of our stories lost until the next rainbow comes. Some find their way back, others become legends. It is all part of the great cycle of water and rest."

"Now go and explore," said Misty happily. "Talk to everyone. Make friends. In Rainbowville we are all one big family."

RAINBOWVILLE
WELCOME!

Aqua waved goodbye and set off into the colorful streets with a heart full of wonder.

RAINBOW CAFE
Be Kind
RAINBOWVILLE STORY SHOP
Stories Bring Us Together

Aqua explored every corner of Rainbowville. The streets were filled with music and color.

Welcome to
RAINBOWVILLE
Drip Drop
CAFE

Every droplet he met had a different story to tell about rivers, rain and faraway places. Aqua listened to every single one.

RAINBOW TREATS
BLOOM & BRIGHT

Even with all the fun around him,

Aqua kept thinking about the clock.

Time was passing.

He wanted to understand
where he came from and
what his purpose was.

He knew he had to keep searching.

Then something wonderful happened!

Aqua bumped into another droplet named Azure. She shimmered like a clear blue sky on a perfect day.

RAINBOWVILLE

Azure had always dreamed of exploring the world beyond Rainbowville. She wanted to see the oceans, rivers and mountains that other droplets talked about.

When she heard Aquas questions, she knew they were meant to explore together.

“Let’s go on an adventure together!” said Azure excitedly.

Aquas eyes lit up.

He still had so many questions swirling inside him, but now he had a friend by his side.

"Yes!" he said. "Let's go!"

RAINBOWVILLE

First, they went to find Rivery. The oldest and wisest droplet in all of Rainbowville. He lived at the edge of town where the rainbow colors faded into soft white mist.

Everyone said he
knew the answers to
everything.
Aqua and Azure walked
side by side.
Their hearts fill of hope,
looking for answers
about where
they came from.

They wanted to understand their place in the great big world of water.

Rivery had big kind eyes and a calm deep voice.

"Ah" he said, nodding slowly.

"You are here because you are ready to learn."

"Ready to learn what?" Asked Aqua.

Rivery smiled, “the journey of water.”

Rivery told them that Rainbowville was born from the magic of the water cycle.

Connecting every corner of the Earth.

Aqua listened closely, then asked “but why us? Why are we here right now?”

Rivery smiled gently.

RAINBOWVILLE

"Sometimes a journey picks you," he said.

"You are here because you are ready to find out who you are and what you were meant to do. The answers are out there waiting for you."

Rivery's stories
filled Azures heart
with excitement.
He spoke of mighty oceans,
rushing rivers and giant
glaciers made of ice.
She wanted to see all of it
with her very own eyes.
She wanted to feel
the water cycle for real!

Together Aqua and Azure went to ask the Council of Clouds for permission to leave Rainbowville.

The Council were ancient cloud being who floated high in the sky and watched over everything with wise, gentle eyes.

The council warned them that the journey beyond the rainbow was dangerous.

Many droplets who had left never came back.

But when they looked into Azures brave eyes, they saw a spark they could not say no to. "You may go," the Council said, "but take this with you."

They passed down
a tiny glowing droplet.
The essence of
Rainbowville itself.
"When you are ready
to come home
this will show you
the way."

Holding the glowing droplet close, Aqua and Azure transformed into mist and floated up to a big white cloud. They were on their way, drifting towards tall mountains and the great wide world beyond the rainbow.

From high up in the clouds they could see everything!

Rivers winding through green valleys
like silver ribbons.

Rain falling softly onto forests below.

The whole water cycle was happening
right in front of them,
and they were part of it!

They laughed and tumbled through the spray of giant waterfalls.

They floated peacefully across the surface of hidden lakes so still they looked like mirrors. Every moment was a brand-new discovery.

The river showed Aqua and Azure what it meant to keep going no matter what. Rivers carve through rocks and stone over thousands of years never giving up.

"Be like a river,"
the water seemed
to whisper.

Always keep moving!

The deep oceans
showed them that
there is always more
beneath the surface
than you can see,
and the slow ancient
glaciers taught them
that even the most
powerful changes can
happen one small
step at a time.

They felt the
Moon gently pull
the ocean tides.

They felt the warm
Sun lift them up
into the sky.
They were learning
that water never
really disappears,
it just changes and moves
and begins the journey
all over again.

As the season changed from warm to cold, Aqua and Azure begin to miss Rainbowville.

They had learned so much about water, but now they understood something even more important; Home is the most precious place of all.

Aqua held up the
glowing droplet
the Council had given them.
It pulsed with warm
rainbow light.
Together they wished
with all their hearts
and a brilliant rainbow
burst across the sky!
Pointing the way back
to Rainbowville.

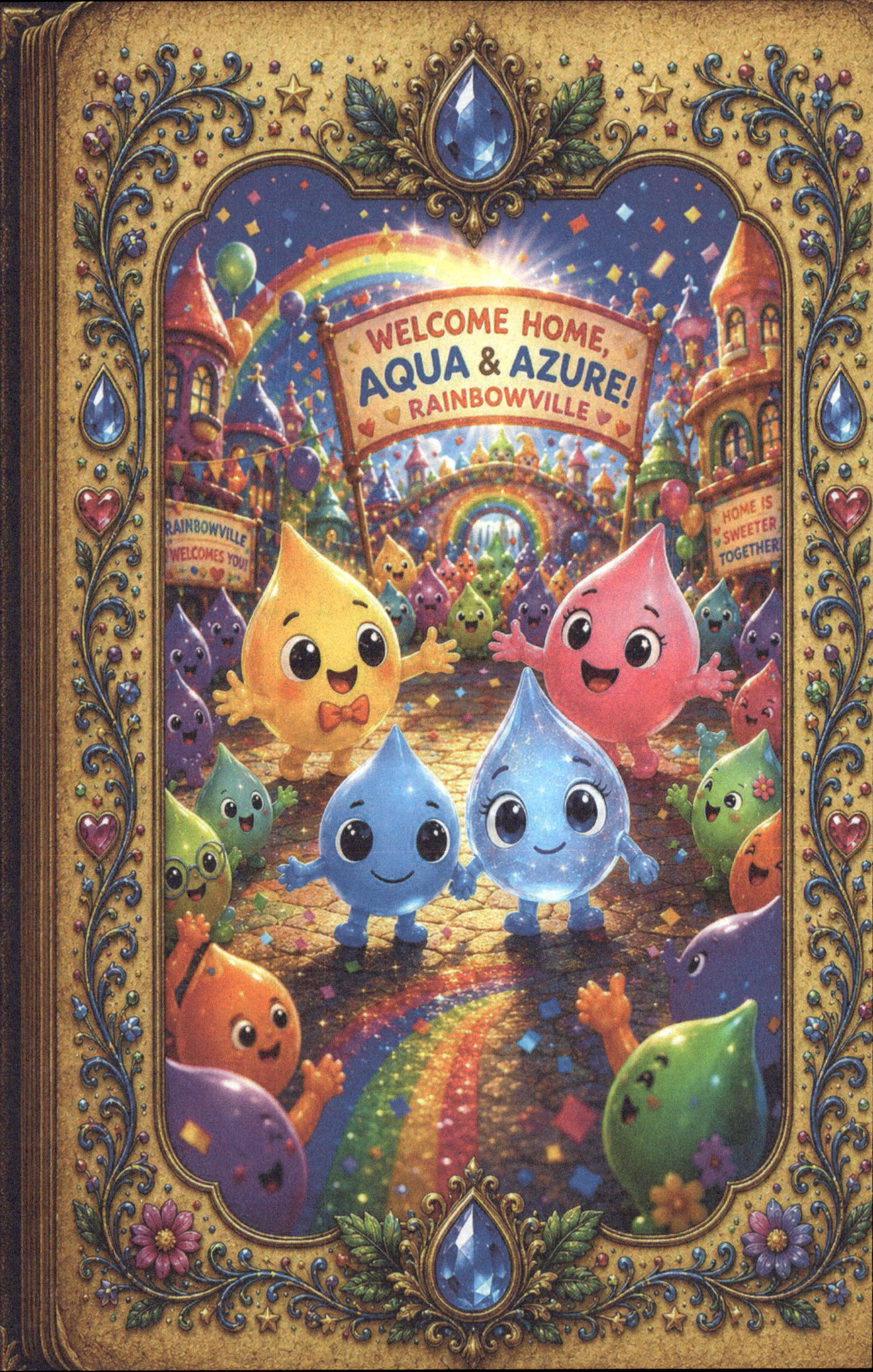
WELCOME HOME,
AQUA & AZURE!
RAINBOWVILLE
RAINBOWVILLE
WELCOMES YOU!
HOME IS
SWEETER
TOGETHER!

When they arrived home the whole town cheered! Mayor Dewdrop and Misty rushed over with biggest smiles. Droplets of every color celebrated. Aqua looked all around Rainbowville and finally felt something he had been looking for. He felt at home.

Aqua and Azure
had become
something special.
They were now the
bridge between
Rainbowville and
the great world
of water outside.

They had seen the whole water cycle and carried its lesson in their hearts forever. And so, the story of Aqua and Azure became part of Rainbowville forever.

It reminded every little droplet that the world is full of wonder.

That every journey teaches you something, and that every single drop of water including you matters.

Speak
Up
speakupboard.com

Made in the USA
Monee, IL
03 November 2026

67384791R00059